iPhone 6 User's Manual: Tips and Tricks to Unleash the Power of Your Smartphone! (includes iOS 8)

By Shelby Johnson

Disclaimer:

This eBook is an unofficial guide for using the Apple iPhone 6 and is not meant to replace any official documentation that came with the device. The information in this guide is meant as recommendations and suggestions, but the author bears no responsibility for any issues arising from improper use of the iPhone 6. The owner of the device is responsible for taking all necessary precautions and measures with their phone.

iPhone 6, iPhone, and Apple are trademarks of Apple or its affiliates. All other trademarks are the property of their respective owners. The author and publishers of this book are not associated with any product or vendor mentioned in this book. Any iPhone 6 screenshots are meant for educational purposes only.

Special Bonus with this Guide

As a special thank you for purchasing this eBook, we are offering "50 Great iOS Apps eBook" for free here at TechMediaSource.com!

Table of Contents

Introduction

As with most technological advances, the iPhone 6 has stepped up its capabilities, improved its design, increased its processing speed and delivered an overall enhanced version of its previous models by leaps and bounds. Both the iPhone 6 and 6 Plus feature upgraded design and specifications over their predecessor, the iPhone 5S, so let's take a look at what's new first.

Among the new updates are the larger 4.7-inch 1334x750 "Retina HD" LCD backlit display with Ion-strengthened glass for the 6 and a 5.5-inch display for the 6 Plus. The 6 features 1334x750 pixel resolution at 326 ppi, while the 6 Plus features 1920x1080 pixel resolution at 401 ppi. Display Zoom and Reachability have also been added on the display screen.

For dimensions, height, width and weight have all increased on the two new iPhone models, while the depth has been trimmed down to make them slimmer. The iPhone 6 is 5.44" x 2.64" x 0.27" with a weight of 4.55 ounces, while the 6 Plus is 6.22" x 3.06" x 0.28" and weighs 6.07 ounces.

The processing power in the new iPhone 6 models has also been upgraded from an A7 chip to an A8 chip with an M8 motion coprocessor. Apple has claimed that this provides 20 percent more CPU power as well as 50 percent more GPU power.

As far as the camera goes, the new 6 and 6 Plus each feature brand new 8-megapixel iSight cameras with Focus Pixels to make photos and videos that much better. Also, the 6 Plus offers Optical image stabilization, a feature that neither the iPhone 6 nor the 5S have on board. Panorama shooting has been improved to handle up to 43 megapixels for both of the new models, something the 5S did not have.

The biggest features on the iPhone 6 and 6 Plus come by way of the mobile operating system, iOS 8. In this guide you'll get insight into how to use all of the aspects of these smartphones with tips and tricks for the iOS 8 as well.

Pre-Setup

Before you buy the iPhone, make sure your cellular provider provides iPhone service. You will also need a broadband Internet service, and a computer that has a USB port, and an operating system of Mac OS X version 10.6.8 or later, or Windows 7, XP. Vista, Service Pack 3 or later. You are also going to need an iTunes account, if you do not have one already.

Sign up for iTunes/App Store/iCloud

Signing up for iTunes/App Store/iCloud capabilities, which you are very much going to need to get the most from your device, you simply need an email address, your choice of a password, and your contact information to set the account up online. You will also need a valid credit card or PayPal account, and will have to include that information in your account set up.

iTunes will not charge your account unless you make a purchase, but it does require a default payment up to date payment plan at all times. This allows you to update apps, download music, movies and iBooks – even if they are free. Your account will ask you to sign in by entering your password before any downloads can take place, and without a valid payment option – even if you are downloading free stuff – you will not be able to proceed with everything your account has to offer.

Once your account is up and running, you can start storing items in your iCloud, so you can access them from any of your Apple devices.

Protect Your Device

Before you set up your iPhone for daily use, you need to invest in a few accessories because this brilliant phone is going to become a true companion going forward. It should also be viewed as an investment; one you want to protect. Purchase a protective case of your choice, whether it is waterproof and shock proof, or just to keep it from exploding should it inadvertently hit the ground. There are literally thousands and thousands of these cases, and they each come in different colors, designs, and capacities, so add a little personal flair if you would like while protecting your device.

In addition to the case, a screen protector is an easy purchase that will help in several ways. First, it will diminish the fingerprints that transfer during use. It will also keep your screen from scratching while it is stored in your purse, briefcase, or backpack. Lastly, find a car charger and keep it in your vehicle. This will allow you to charge your device while you travel, so you are never left without its functionality.

Export Contacts from Old Phone

Depending on what your old phone is, and for the purposes of this guide, the assumption will be made that you already have a smartphone (no matter the brand). If so, count on your current cellular provider to help by using the online tools they provide.

Verizon

- Register for Backup Assistant.
- Install the Backup Assistant App on your Current Phone (the one with the contacts) through the App Store.

- Load Backup Assistant, Log in, and Back up Your Contacts.
- On your iPhone 6, open the App Store and install VZ Contact Transfer. Start the program and follow the onscreen prompts.

Sprint, AT&T and T-Mobile

- Install & Run Google Sync on your Current Smartphone.
- Install & Run Google Sync on your new iPhone 6.

This will require a Gmail account to accomplish, and if you do not already have one it will only take seconds to register for one that you will use for your new device going forward.

What's in the iPhone 6 Box

When you purchase your new, sleek and exciting iPhone 6, and carefully open the heavy-duty packaging, you will find:

Headphones with a Built-in Microphone: These headphones work just like every other pair of ear buds you have shoved into your ears before, but this version comes with a microphone on the cord that lands right near your mouth, so you can talk using the hands-free approach, and hear the other person through the ear buds. It also comes with a storage and travel case, which may help you hang onto them a little longer than the last set you had.

Lightning USB Cable: The cable has a standard USB output on one end, which will fit directly into your computer or other device that has a USB input. The "lightning" end is a smaller input for more efficient charging and transfer, and is not compatible with the previous 30-pin cable the previous iPhones use.

Charging AC Power Adaptor: The charger will plug directly into a standard AC power outlet, while the USB cable attaches to base of the adapter on one end and into the phone on the other.

Paperwork: Included in the box is a bundle of paperwork including Apple stickers, warranty details, and a diagram of the iPhone 6.

Setting up iPhone 6 Out of the Box

First things first, when you take all of the items out of the box, you are going to have to charge the phone fully before you start using it. You can plug the lightning adaptor into the phone and attach it to the wall adaptor or directly to your computer (in a USB port) and allow it to charge fully.

Once you take the iPhone 6 out of the box, and get it charged up, you will be ready to set up your iPhone 6 and start using it to talk, text, surf the web, listen to music, and any of the other number of other amazing things that this phone can do.

Adding the SIM Card

The iPhone 6 uses a nano SIM card, which you must insert before setting up your phone (unless your phone already came with the SIM card inserted). Using either a special tool or a paper clip, insert the tip into the small hole on the SIM card tray to eject it (like a very tiny CD player tray would, except that it comes all of the way out). Place the nano-SIM inside the tray, and press it back into place, flush with the device. There's a great visual walkthrough on how to do this at the YouTube video here.

Initial Set Up Screens

From the moment you turn your iPhone 6 on, which happens automatically when you plug the device in, you will receive a step-by-step tutorial on setting up your phone.

First you swipe to the right at the bottom of the phone. From there, you choose your language. Next, you select your country or region. Finally, you choose a Wi-Fi network to connect to.

To join your Wi-Fi network, select it from the list provided, then type in your password, and tap "join."

The news step is to choose to either enable or disable location services. The location services setting allows apps to use your approximate location to help provide you with accurate data. Tap either "Enable Location Service" or "Disable Location Services."

Now it is time to set up your iPhone. You have 3 options from which to choose for the setup. You can choose "Set up as New iPhone," "Restore from iCloud Backup," "Restore from iTunes Backup."

‹ Back

Set Up iPhone

Set Up as New iPhone ›

Restore from iCloud Backup ›

Restore from iTunes Backup ›

What does restoring do?

Your personal data and purchased content
will appear on your device, automatically.

Make your selection depending on if you have or have not
used a previous iPhone and if you did or did not back it up
somewhere.

After making your selection, you will be asked to either sign
in with your Apple ID or to Create a Free Apple ID. If you do
not have an Apple ID, tap "Create a Free Apple ID," otherwise
simply tap "Sign In with Your Apple ID."

Next sign in with your Apple ID. Simply enter your Apple ID and the Password. Upon signing in, you have to agree to the Terms and Conditions. There is an option to have them sent to an email address, so if you would like, you can send them to your email. Tap "Agree." When the screen pops up tap "Agree" again.

After a few moments, your phone will arrive on the iCloud screen. The iCloud option allows you to access your photos, videos, and other stuff on all your Apple devices. Choose to either use iCloud or not use iCloud depending on your personal preference.

Next you will decide if you want to use the Find My Phone feature. This feature allows you to locate, lock, and erase your iPhone 6 in the event it is lost or stolen. Choose either "Use Find my iPhone" or "Don't Use Find My iPhone." If you choose this service, once again you will be prompted to agree to the terms of service. Tap "Agree" and tap it again when the screen pops up.

Next you have the option to set up Touch ID. This allows you to use your fingerprint instead of a passcode and Apple ID for any purchased you make. If you would rather not set up this feature tap "Set Up Touch ID Later." To set up Touch ID, follow the prompts and touch your thumb or finger to the Home button repeatedly. You will feel a slight vibration in the phone each time you are to pick up your digit and place it on the key again. Finally, once the iPhone 6 has mapped your fingerprint, you will be asked to create a 4-digit passcode. Enter the 4 numbers and then reenter them. Next you have the choice to use Touch ID for iTunes and the iStore. You can choose to use it then or choose to set it up later.

The next screen is the iCloud Keychain. The iCloud Keychain keeps all your passwords and credit card information up to date on all your approved Apple devices. Tap either "Set Up iCloud Keychain" or "Set Up Later." To set up the iCloud Keychain, you will be asked to use the passcode you just created or to create a new one. Tap either "Use Passcode" or "Create Different Code" depending on your preference.

Finally, you will be prompted to enter your Phone Number. Simply type in the Country Code (i.e. 1 for the United States), and enter your phone number in the space provided.

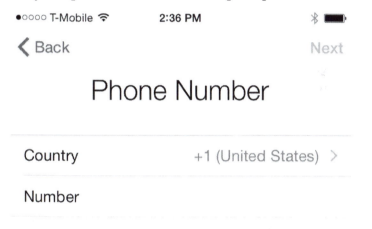

Next you will decide to Use Siri or to not use Siri. Siri is Apple's virtual assistant, and she can help you make a phone call, find a restaurant, or get directions. Tap either "Use Siri" or "Don't Use Siri."

Siri

Use Siri   >

Don't use Siri   >

What is Siri?

Siri helps you get things done just by asking.
You can make a phone call, send a
message, dictate a note, or even find a
restaurant.

Siri sends information like your voice input,
contacts, and location to Apple to process
your requests.

About Siri

The Diagnostics screen is next. Here you will choose to "Automatically Send" or "Don't Send." The Diagnostics helps Apple improve its products, and it can include your location, so this is completely up to you.

Next is App Analytics. This allows app developers to improve their apps for the iPhone 6 by allowing Apple to share data like apps crashing and other app statistics. Tap "Share with App Developers" or "Don't Share."

Display Zoom is the next feature you will setup for the iPhone 6. Here you can choose the standard view or a zoomed view depending on your needs. No matter which view you choose, you can change it later in the Settings should you need the other view.

Finally, your setup is complete Welcome to iPhone!

●●○○○ T-Mobile 📶   2:41 PM   ⚹ ▬▬▬

Welcome to iPhone

Get Started

Import or Add Other Phone/Contacts to iPhone 6

There are a few ways to sync your contacts using iCloud or Google Contacts. If you currently have an Apple device, you should already have an iCloud account. If not, you can create one by going to the settings tab on your iPhone 6 and tapping iCloud. You will receive a free email address, and five gigabytes of storage.

As long as the iCloud was set up on your previous device, you can simply download everything in it to your new iPhone 6 effortlessly. To be sure all of your information stays backed up – and current on all of your devices, turn on iCloud backups by tapping:

1. Settings.
2. iCloud.
3. Log in with your Apple ID and password.
4. Go to Storage & Backup.
5. Turn on iCloud Backup.

You can also sync to your iTunes, simply by plugging the phone into your computer, opening iTunes and tapping "Sync" in the program.

Use Google Contacts to Sync your Phone's Contacts

If you are transferring from a previous smartphone to an iPhone for the first time, you can use Google Contacts just as easily, and here's how:

* Install and run Google Sync on your current phone.
* Install and run Google Sync on your iPhone 6.

How to Export a .vcf File and Import to iPhone

If you do not have a smartphone, you are going to have to work a little harder to get your contacts transferred, but it will be worth it.

First you are going to have to create a backup file with your existing carrier by following the following steps.

Verizon

1. Register for Backup Assistant.
2. Install the Backup Assistant App on your Current Phone (the one with the contacts) through the App Store.
3. Load Backup Assistant, Log in, and Back up Your Contacts.
4. On your iPhone 6, open the App Store and install VZ Contact Transfer.
5. Follow the Instruction that Appear Onscreen.

Sprint

1. Register for Sprint Mobile Sync.
2. Activate Sprint Mobile Sync on Your Current Phone.
3. (Main Menu > Settings > Contacts > Mobile Sync > Activate).
4. Log into Sprint Mobile Sync from your home/office computer; export your contacts as a .CSV File.
5. Log into Gmail and import the .CSV file.
6. Install and Run Google Sync on your iPhone 6.

AT&T

1. Register for AT&T Mobile Back Up on their Website.
2. (You will receive a link on your current phone to install the app).
3. Load the Mobile Back Up App, and Back Up Your Contacts (it is a presented option).
4. Log into AT&T Mobile Back Up from your home/office computer; export your contacts as a .CSV File.
5. Log into Gmail and import the .CSV file.
6. Install and Run Google Sync on your iPhone 6.

T-Mobile

1. Your T-Mobile phone comes loaded with a "Mobile Back Up" app.
2. Load the App and Select the "One Time Sync" Option.
3. Log into T-Mobile Mobile Back Up from your home/office computer; export your contacts as a .CSV File.
4. Log into Gmail and import the .CSV file.
5. Install and Run Google Sync on your iPhone 6.

Spring Cleaning App

Chances are, you have a few duplicate contacts on your phone, or some that are even incomplete or used less than others. The free Spring Cleaning app was developed specifically for Apple devices, and allows the user to:

• Perform Easy Toggling Capabilities between Selected and All.
• Empty the Trash Bin Permanently.
• Restore Deleted Contacts.
• Search Contacts and Trash for People.

- Select and Delete Multiple Contacts Effortlessly.
- View Full Contact Details.

Setting up your Email, Facebook and Twitter Accounts

Of course you will want to set up all your social media like Facebook and Twitter as well as your email on the iPhone 6. The following sections explain how to do each one.

Email

To set up your email account so you receive the messages directly on your iPhone 6:

1. Tap Settings.
2. Locate & Select "Mail, Contacts, Calendars."
3. Select "Add Account."
4. Choose from the list of accounts you can add.
5. Enter your account information, and tap "Next."

Other

The accounts from which you can choose are all the popular options including iCloud, Gmail, Yahoo, and Outlook to name a few, as well as an "Other" option that will allow you to manually enter your account information.

Facebook

To set up Facebook on your iPhone 6, first you will need a Facebook account, if you do not already have one. This is typically easier to set up on your desktop or laptop computer by going to Facebook.com and following the steps to create an account and a profile.

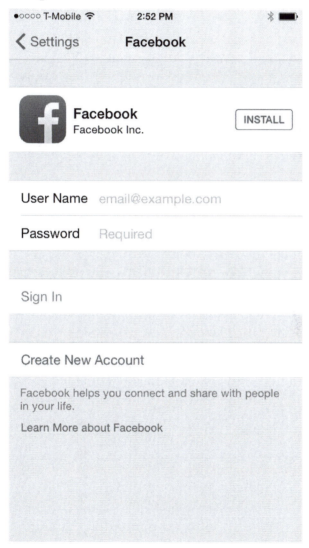

When your account is set up, or if you have an existing account, complete the following to set up Facebook on your iPhone 6:

1. Tap Settings.
2. Locate & Select "Facebook."
3. Enter your Facebook User Name and Password then tap Sign In.
4. You will be asked if you want to install Facebook now, or wait until later. If you choose to install, your phone will connect to the App Store and begin downloading the Facebook app. If you set them up, you may be prompted to enter your passcode or Touch ID.

Twitter

The same goes for Twitter. If you do not have an account, you will need one.

1. Tap Settings.
2. Locate & Select "Twitter."
3. Enter your Twitter User Name and Password then tap Sign In.
4. You will be asked if you want to install Twitter now, or wait until later. If you choose to install, your phone will connect to the App Store and begin downloading the Twitter app. If you set them up, you may be prompted to enter your passcode or Touch ID.

Setting Up Security on the iPhone 6

When you want to keep the contents of your iPhone 6 to yourself, or if you simply have children who are always tapping away at your screen, you can assign a passcode to the device. If you did not choose a passcode earlier, you might want to now.

You will have to enter this passcode every time you access the device, even to check email or make a phone call. To set a passcode, simply:

1. Tap the Settings Icon.
2. Locate & Select "Touch ID & Passcode" from the Menu.
3. Choose to set a Passcode or Add a Fingerprint.
4. Follow the on screen instructions for whichever choice you made.
   Note: *If you choose to add a fingerprint, you will also have to create a 4-digit passcode.*

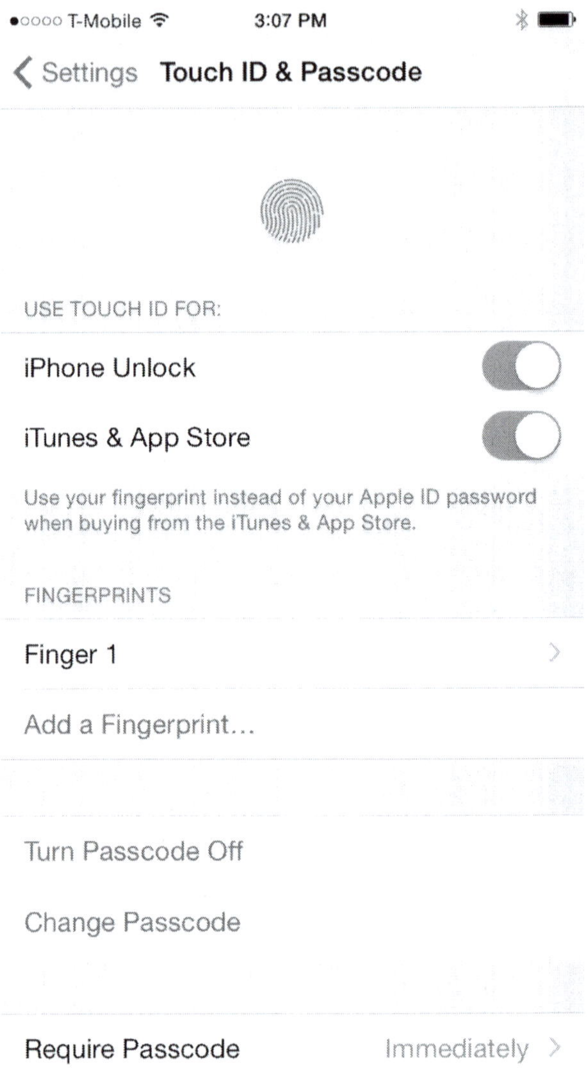
●oooo T-Mobile
3:07 PM
Settings
Touch ID & Passcode
USE TOUCH ID FOR:
iPhone Unlock
iTunes & App Store
Use your fingerprint instead of your Apple ID password when buying from the iTunes & App Store.
FINGERPRINTS
Finger 1
Add a Fingerprint...
Turn Passcode Off
Change Passcode
Require Passcode
Immediately

From this Touch ID & Passcode screen you can also choose whether or not to use the passcode for the iPhone Unlock and the iTunes & App store as well as turn the passcode off completely or change the passcode.

If you do not like the simple passcode, you can also turn that off and require a longer passcode for your iPhone 6. Finally, if you really want to make your phone extra secure you can turn on "Erase Data," which will erase all the data on the phone after 10 failed passcode attempts.

In this screen, you will also choose which items can be accessed from your iPhone 6-lock screen including Today, Notifications View, Siri, Passbook, and Reply with Message. If you want to turn any of these off, simply touch the toggle and slide it to the left.

Getting to Know iPhone 6 Basics

Your new iPhone 6 may take some getting used to if it's your first one. Otherwise, you may be ok to skip this section to check out other helpful information.

Here are some of the basic iPhone 6 functions you'll want to know:

The "Home" button is the circle on the bottom of your iPhone 6 with a square inside it. Pressing this button will usually take you out of whichever app, game or other screen you're currently on, and to your iPhone's main screen.

Top display bar – At the top of your iPhone 6 screen you'll always see a thin display bar. On the left side, it will show your 3G or 4G network, and your Wi-Fi signal strength (if enabled).

●○○○○ T-Mobile 📶   2:41 PM   ＊ ▭

In the center will be the current time of day based on where you are geographically. On the far right will be your current power display, which shows as a battery bar that will diminish as your phone loses battery power.

App icons – Your iPhone 6 home screen is full of different icons for each of the apps that your iPhone included as well as any that you have downloaded (free or purchased). Among those that are included on the main home screen are Messages, Calendar, Photos, Camera, Videos, Maps, Weather, Passbook, Notes, Reminders, Clock, Stocks, Newsstand, iTunes, App Store, Game Center and Settings.

Some of these icons may display a red circle or oval on their upper right corner with a number inside. This lets you know that there are new updates, messages, or emails for you to read, depending on the app's purpose.

Settings – The settings app icon is located on your Home screen usually towards the very bottom of the display.

By tapping this particular button, you have access to all sorts of additional settings which include: Airplane Mode, Wi-Fi network, Bluetooth, "Do Not Disturb," Notifications, General, Sounds, Brightness & Wallpaper, Privacy, iCloud, Mail, Notes, Contacts, Reminders, Phone, Messages, FaceTime, Maps, Safari, iTunes, Music, Videos, Photos, Camera, iBooks, Podcasts, Twitter, Facebook and several other features such as Nike + iPod (if you use this function).

Lower panel of icons – On the very bottom of your display will be four icons for different apps: Phone (which is to use the phone function of the iPhone), Mail (for checking your latest email messages), Safari (your standard web browser) and Music (for quickly accessing and playing your latest digital music files).

Notifications – Many of the installed apps will send notifications to you on your iPhone 6. For example, if you get a new Facebook message or comment, or some sort of text message. Also, several apps may display their latest info here, including weather, stocks or other apps you've enabled.

To see your most recent notifications, you can press down where you see the current time at the top of your screen and drag your finger down to bring down all of your notifications. Once here, you can choose to clear out any notifications, or press on them to see what they are about.

Control Center – The Control Center gives you instant access to the camera, calculator, AirPlay, and other features like Wi-Fi, Bluetooth, and Airplane Mode. You can also adjust the brightness, lock the screen in portrait orientation, turn wireless services on or off, and turn on AirDrop.

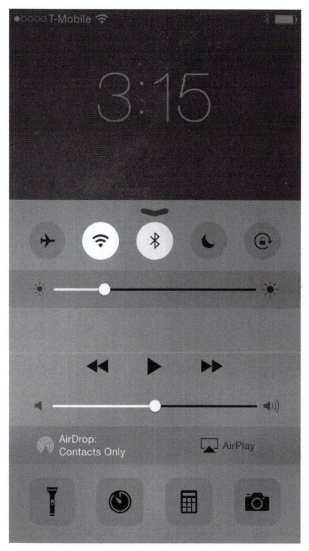

To access the Control Center, simply swipe up from the bottom of any page.

Using the iPhone 6 as a Phone

The iPhone 6 is so many things, but first it is a phone. The following sections explain the phone's phone features and how to use them.

Making Phone Calls

Making a phone call is the most basic use of a smartphone, and the iPhone 6 handles that task quite easily. To place a call you'll just press the green phone icon, which is located on the lower left of your home screen display. Once you've done this, you can choose from your contacts to place a call to one, or go to the "keypad" option seen on the lower panel of options on the display.

Use the keypad to enter the number you want to dial and press the green phone "Call" button at the lower center of screen display to make the call.

Favorites – Facetime

The Favorites portion on the lower phone display is denoted by a "star." This option allows you to store certain individuals' email or phone numbers for using the FaceTime option with. Keep in mind, this is going to only work with other users who use FaceTime, so you'll want to store individuals there who have an iPhone or iPad or other Mac OS device that has the capability for the face-to-face video conferencing.

To add a favorite, simply tap on the star icon at lower left of phone display screen. You can then click on the "+" symbol at the upper right of your display to add a new FaceTime favorite from your list of contacts.

Recents Screen

In the lower part of your phone display screen you'll see "Recents" denoted by a clock symbol. Tapping on this symbol brings up your recent call or FaceTime contact history. You can tap on the arrow pointing to the left next to any of the names of people or places you called and that will bring up a new screen showing you when the call was made including what date, time and the duration of the call. The screen also offers you the options to send a text message, start a FaceTime session, share the contact or add to your favorites.

Contacts & Keypad Screens

Contacts were covered in a previous section, but it's worth noting you can tap this particular option and add new contacts, or simply select a previously entered contact to call up or send a text message to.

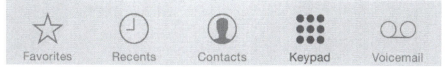

| Favorites | Recents | Contacts | Keypad | Voicemail |

The keypad feature is self-explanatory, as this gives you a keypad to tap in a phone number and place a call. Additionally, you'll notice a small "+" with a person symbol on the lower left portion of the keypad. Tapping on this lets you Create a New Contact or add the Number you're entering to an existing contact in your iPhone.

Emergency Call

If you have a passcode set on your iPhone, there's a convenient Emergency call feature. This will allow you to bypass the code entry and make a phone call fast if needed.

To access the feature, from the passcode entry screen tap the "Emergency Call" button located on the lower left part of the keypad on your display. It will take you to the screen shown above, where you can now tap in a phone number and then hit the green phone button to make your call.

Tip: *For better reception, try switching off LTE as it could give you better coverage in certain regions. To do this, go to Settings > General > Cellular and where it says "Enable LTE" tap the button to turn OFF. This may or may not help improve call reception in certain regions. Keep in mind you'll want to turn that back on if you want to take advantage of the high speed LTE networks again.*

Using Important iPhone 6 Features

Your new iPhone 6 has a ton of really great features, and you are going to want to use them to their best of your ability to get the most out of your device.

How to Use Fingerprint Sensor

The iPhone 6 and 6 Plus phones have a unique feature that first came along with the iPhone 5S device. The fingerprint sensor on these phones provides extra security for owners of the device so they can unlock their device using the press of a fingerprint, rather than a passcode. This is referred to as Touch ID.

Here's how to set up and use the Touch ID feature:

1. Tap on the Settings icon.
2. Tap on "General" option.
3. Tap on "Passcode & Fingerprint" option.
4. Enter your passcode if you currently have one.
5. Tap on "Fingerprints" and then tap on "Add a Fingerprint."
6. Next, you'll be instructed to place your finger on the circular home button (with fingerprint facing it) on the bottom of your phone.
7. You'll need to "lift and rest" your finger on the home button repeatedly, until a proper fingerprint has been generated on your screen. The display will prompt you with the necessary instructions as you go until "Success" is displayed on your screen.

Once you've set up the fingerprint sensor, it's very easy to use. Now when you go to use your phone and it's been sleeping, you can simply rest your finger on the home button and the iPhone will become active for use.

Note: *A quick tip, you can set up multiple fingerprints for the Touch ID, such as more than one finger from your hands, or someone else's finger who may need frequent access to the phone.*

TIP: *Some 6 or 6 Plus owners might be concerned that if they lose their phone, a good Samaritan can't make a call to return the phone. There may still be the ability to make "Emergency" phone calls using the iPhone without the fingerprint security feature being activated. Simply slide to unlock to get to the passcode screen and you'll see the "Emergency" option at the bottom left corner of the display which you can tap to make a call.*

How to Use Siri

Siri is a personal assistant that can be summoned to do "chores" for you, simply by asking her. Siri requires Internet access and can be awakened by simply holding down the "home" button – the main button of your device, at the bottom that contains a square – for a second. You will hear a two beeps, and Siri asking how she can help. You can ask her for the nearest Chinese restaurant location, or to call a friend who is listed in your contacts by simply saying, "Call Sarah Marshall." She understands commands, so you do not have to talk like a robot to get her to respond accordingly.

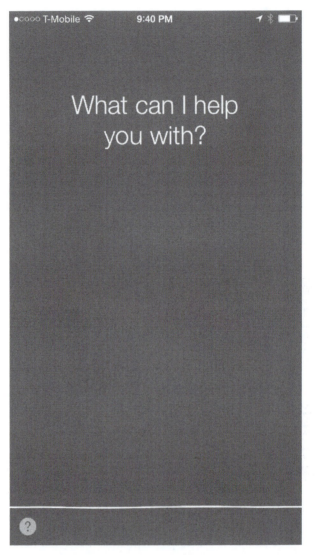

What can I help
you with?

She can also compose a text message or read any messages in your inbox to you aloud. You can interact on social media just by telling her which platform to access, and typing the message you wish to display. She can also find a plumber, and dial their number if you command her to. She will set your alarm, set a timer, and add items to your calendar.

Siri will read your requests back to you, or ask for clarification if she could not understand you. Use her for everything, even weather updates in different cities! She never grows tired of your requests, and is usually very accurate with her responses.

Siri Settings & How to Disable Siri

You can customize your Siri for your iPhone 6, simply by going to Settings > General > Siri on your phone.

In this settings area, you can also adjust certain aspects of Siri, such as changing your language, or making it so that when you raise your iPhone to your ear, Siri automatically is activated for you. This could be a helpful feature to enable because you can raise the phone to your ear, and then request Siri to make a call to certain contacts you have stored.

Should you decide you don't want the Siri feature active, you can always disable this feature. Simply go to Settings > General > Siri where you can switch the ON button to OFF.

How to use Voice to Text

One of the great features that is part of the iPhone 6 is Voice to Text. This helpful component of the smartphone allows you to speak to the phone and have it make your spoken words into text on the screen. This will work with a wide number of apps including text messages, e-mails, Google searches, reminder notes and many other applications on the phone.

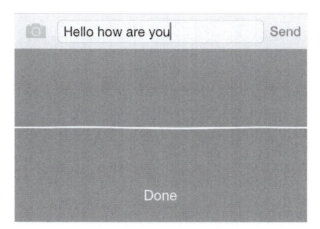

To activate Voice to Text, look for the microphone symbol that appears on the lower left area of your on-screen keyboard. Tap on the microphone and wait for a double chime sound, then speak the words you want transcribed.

Keep in mind, there may be a learning curve and the iPhone may not pick up every word perfectly. It also uses the Internet at times to search for various words, so if you've got a bad or low signal Wi-Fi or network connection, it may have trouble.

How to use Search

You may at some point want to find a file, contact or other piece of information on your iPhone 6. The smartphone has a built in feature to help you quickly find things on your phone. It will search through pretty much everything on your iPhone - text messages, emails, contacts, digital music, photos, and other files.

To get to the search screen, simply swipe/drag your finger down from any screen. This will bring up the Search screen that allows you to type in (or speak with voice-to-text) a name of a file, person or other tidbit of info. As you start typing, suggestions will pop up that may help you. This is a great way to locate files, names of people, or other info on your phone.

If you have a lot of music on your phone, you might type in "Prince," for example, if you want to bring up all the instances of Prince songs or files on your iPhone. You can also type in your search word or phrase and scroll down on the screen to tap on "Search Web" or "Search Wikipedia" to search these particular resources for more information on a term or phrase.

How to Update iOS

The operating system for Apple iPhone 6 is iOS 8, and it controls everything your phone does. When an update is available, your phone will alert you to its readiness so you are operating on the most up to date system that was designed for your device. It is what allows you to operate the phone as a whole, much like Windows 8 for your computer. If you skip an update manually when asked, you can always go back to it later and download it when you are ready by following these steps:

1. Tap Settings.
2. Tap General Tab.
3. Tap Second Option "Software Update."

It will reveal the last update you skipped, and all you have to do is tap the option to download it. You will not be able to use your phone for a few minutes while it downloads the update.

How to use iPhone Camera for Photo & Video

Your iPhone 6 camera takes beautiful pictures and video with a stunning new eight-megapixel iSight lens that will record 1080p video at 60 frames-per-second, or slo-mo video at 240 frames-per-second. Panoramic, time-lapse, and a photo timer also come in very handy. Basically, you can easily record anything and everything you desire, with a simple click or two.

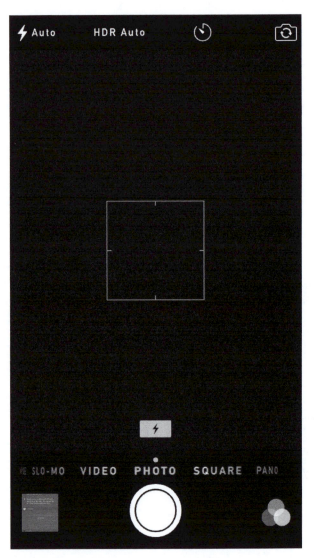

To access the camera for pictures or video:

1. Locate & Select the "Camera" App that is Pre-loaded on the iPhone.
2. You can zoom by using your fingers to push the frame in or out (which will bring up a Zoom bar on the bottom of the screen for easy use).
3. Other options are available to adjust your image including the "On/Off" button at the top left of the screen to control

flash, the "Options" button to control HDR, Grid or Panorama modes, and the Reverse Camera button. This will switch from your iPhone's front-facing to rear-facing camera, or vice-versa.

4. To take a picture, simply tap the "Camera" icon located on the display screen (it is a camera inside of a circle) - or you can use the "+" volume rocker button on the outside of your phone to make it easier.

5. The Picture will automatically save to Your Camera Roll for later use.

Note: *You can also tap on the small photo in the bottom corner of your screen to bring up your latest image, and then choose various options such as to send it to Facebook, a text message, email, a printer or use it as wallpaper on your phone.*

For video recording capability simply toggle the button on screen to the video recording icon, and proceed as you would with taking a picture. It will automatically save to the phone.

How to use Facetime

FaceTime is a great way to keep in touch with other people who have an iOS device. If your best friend is 7000 miles away and has an iPhone, iPad or iTouch, you can contact him or her via FaceTime and literally talk face to face over your Internet connection.

Simply tap on the FaceTime icon that is pre-loaded onto your iPhone 6, and tap "Contacts" at the bottom of the screen. Locate the person you wish to reach, and tap their contact information.

Once you do, you will hear a ringing sound – and so will they. When he or she answers, your screen will be filled with their face and surroundings, and their device with yours. You will see yourself in a smaller screen in the corner of the phone's display.

You can talk just as you would during a phone call, as there is no need to shout. You can also move the device around, or swap the camera to the outside lens so the other person sees your surroundings, instead of you. This is great way to show them your apartment, the landscaping around your house, or to let them wave at everyone else in the room.

How to use Find my iPhone

Find my iPhone is a free app that you can download to your iPhone 6. When you do, it will ask you to register the device – and others, if you have another like an iTouch or iPad – so you can see where it is at all times. Once you have registered the device, be sure that it is turned on in your iCloud section, so you can actually use it.

1. Tap Settings.
2. Locate & Tap iCloud.
3. Locate "Find My iPhone."
4. Slide the Selector to "On."

When you cannot locate your iPhone, simply logon to your iCloud on your desktop or laptop, and locate the "Find My iPhone" section. When you do, you will be given several options:

Play Sound: This will command your phone to play a sound for two minutes, so you can locate it in your home, car or office – assuming it is near you to begin with. This "sound" is more like a siren than it is a beautiful medley, so you will find it rather quickly if you have misplaced it.

Lost Mode: If you cannot find the iPhone anywhere you are, this command will allow you to lock it and display a contact phone number for another person to call when they find it – assuming they are nice enough to return it. When you get it back, you will be able to tell exactly where the phone was through the app's tracing capability.

Erase iPhone: If you are certain you lost your iPhone, and want to protect your privacy, you can erase all of your personal information, music, accounts, etc., remotely with this command. It will restore the iPhone to its factory settings.

How to Use iCloud Across Devices

iCloud is virtual storage, five gigabytes of which comes free with your account, which is quite a bit of space to enjoy. It allows you to store music, photos, videos, calendars, contacts, and all of the iPhone's supported documents virtually, so they are being held off – or outside of – the actual device. This means that if you lose your iPhone, all of your information exists in the iCloud, so you can download it to your new, replacement device effortlessly.

☁ iCloud Drive	Off >
✿ Photos	Off >
✉ Mail	◯
👥 Contacts	◯
📅 Calendars	◯
☰ Reminders	◯
🧭 Safari	◯
📝 Notes	◯
🎫 Passbook	◯
⟲ Backup	Off >
🔑 Keychain	On >
◉ Find My iPhone	On >

ADVANCED

If you have more than one Apple device, like an iPhone and an iPad, the iCloud will allow each device to share the same information. So if you take a picture with one, it will add it to the other via the iCloud. The same goes for contacts, changes you make to your calendar and everything you purchase along the way.

The good news is your purchased music, apps, books, and television shows do not count against your storage allotment, so five gigabytes should be plenty of space, although you can always purchase more effortlessly.

Sign up for an iCloud account, if you have not already by going to icloud.com or tapping the option under the settings icon on your actual device. Once you do, you will be given the option to share your information between devices, by simply pushing the appropriate slider to "on" or "off". The device will back up on its own each day, or you can simply press the "Storage and Backup" button and hit "Backup Now" to do so immediately.

How to use Maps for Navigation/Directions

A map application comes standard on your iPhone 6, and the icon is simply a close up detail of map directions, with the word "Maps" below it. Tap the icon to search locations, and zoom in directly on everything you are looking for, whether it is the Pyramids in Egypt (because you are curious), your childhood home (still curious), or an actual destination you are pursuing.

You can also get directions by typing a "Start" location – which by default is your current location – and your end destination. The map will drop a pin on your destination, so you can watch as you get closer to your arrival – while keeping an eye to make sure you are actually getting closer.

Users may also download their own preference of directional maps, including Google Maps app, which is a preferred directional app for a lot of people.

How to use Safari Browser

When you are ready to search the Internet for anything and everything, Safari comes standard with the iPhone 6.

Safari works just like any other browser in that it allows you to type a website URL directly into the toolbar to search all of your favorite sites effortlessly. In fact, when you are ready to bookmark all of your favorites, you will find five icons on the left hand side of the URL entry that will help you navigate the page you are on for saving, forwarding or adding to your social media outlets.

- Left Arrow: Allows you to move backwards to a previous page.
- Right Arrow: Allows you to move forward to a different page.

- Opened Book: Allows you to bookmark the page you are on, view your search history or list a page for future reading.

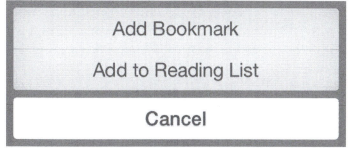

- Cloud: Allows you to view all of the opened tabs on your other devices that are connected through the iCloud.
- Arrow Shooting from a Box: Allows you to email, message, Tweet, or Facebook the content directly. You can also copy the page, add it to your home screen, print, bookmark or add to your reading list from this icon as well.

How to use Other Browsers

You can download other browsers through the App Store, including Google Chrome, Dolphin and Atomic Web. If you prefer to use a browser besides Safari you are welcome to download it. However you cannot delete Safari as a result. It will stay with your device as a default browser.

How to Print from iPhone 6

You will have the option to print from your iPhone 6 with a built in option call AirPrint. AirPrint locates and connects to a printer in your home (hotel or business) wirelessly as long as the printer has the capability to do so.

To print, you tap on a small box with an up arrow icon, located at the bottom of the page. This brings up multiple options including Airdrop, eMail and print. To print, simply tap "print." Not all printers have this capability, but if they do you will be able to connect to it from your phone, and print wirelessly.

How to Add Apps to iPhone 6

Apps are fun, simple and easy to download to your iPhone, and just as easy to delete, as long as they are not factory issued (like Safari, Maps, etc.).

Simply tap on the App Store icon and search for apps by the most popular, featured apps or type in what you are looking for in the search bar at the top.

Apps & Games
for iPhone 6

Best New Apps   See All ›

Hulu Plus	Strava Running	NBA Game Time	Marco
Entertainment	and Cycling -...	2014-15	Weatl
FREE	Health & Fitness	Sports	Educa
	FREE	FREE	FREE

Best New Games   See All ›

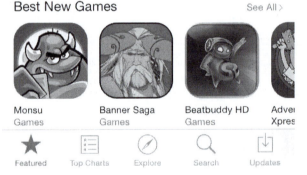

| Monsu | Banner Saga | Beatbuddy HD | Adver |
| Games | Games | Games | Xpres |

★   ☰   ⊘   🔍   ⬇
Featured   Top Charts   Explore   Search   Updates

You will see returns for just about everything you could imagine, and when you have found it tap the word "free" or "(price)" to begin the process. Once you do, if it is free you will have to tap the button again that says "Install App." If it is a pay app, you will have to agree to pay by hitting "Buy App" and then signing into your iTunes account for verification. The credit card or PayPal account you have on file will be charged directly for your purchase.

The app will download automatically, and once it has you are free to use it as you see fit. If you want to delete it, simply hold down the icon until it begins to vibrate and exposes an "x" in the corner. Tap the "x" and you will receive a notice asking if you are sure you want to delete the app. Say yes, and it will go delete it automatically.

Other Music Apps

If you have a free or paid subscription to Spotify or Pandora, you can easily access these two music options simply by downloading their apps to your iPhone. Tap on the App Store and search for the version you prefer before downloading it. If you do not currently subscribe to either one, you can do so for free.

Pandora

Pandora is a selection of radio stations that make suggestions based on the artist that you input. If you list Nirvana as your preference, you will also hear Foo Fighters, Hole and Pearl Jam. The Genre of the artist you enter will resonate for as long as you are listening to that selection. You can certainly change it to Jimmy Buffet radio or Otis Redding radio, as you see fit. When you listen for free, there are ads that play in between the songs, much like a regular radio station, and you are limited to 40 hours of listening per month on your mobile device. The paid service is $36 per year, and allows for unlimited listening, ad free.

Spotify

Spotify requires a Facebook account to access, and provides you with six months of free listening with your trial account, although it does include radio style advertising. You can search music by genre, artist, title and even album. Once the free trial is up, you can keep the service without paying, but listening is limited to ten hours per month, in 2.5 increments per week. Although unused hours carry over, you can purchase a membership for unlimited usage $8.70 per month.

How to get Music, Books & Other Files on iPhone 6

Your iPhone has a lot of capabilities, including the option to listen to music, read books or even watch your favorite television programming or movies directly from your device. There are a number of ways to download this material for purchase and for free. The good news is, unless you are renting a movie or checking out a book from a library, everything you download is yours to keep.

iTunes Store

The iTunes Store icon is automatically loaded onto your iPhone when you purchase it. If you have an iTunes account already, you can access all of the music, movies and books you already own simply by logging onto your account.

Genres   [Featured | Charts]   ☰

U2

**U2's new album is exclusively on iTunes.
And it's a gift to you.**

New Releases   See All >

U2	747	Prince	PLEC

Songs of Innocence
U2

747 (Deluxe)
Lady Antebellum

ART OFFICIAL AGE
Prince

PLEC CTRL
Princ

iTunes Festival London 2014

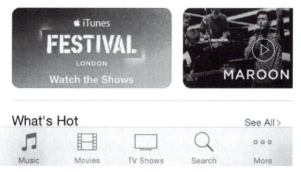

🍎 iTunes
FESTIVAL
LONDON
Watch the Shows

MAROON

What's Hot   See All >

🎵 Music   🎞 Movies   📺 TV Shows   🔍 Search   ○○○ More

Buying new material is just as easy. Tap on your iTunes icon, and search for items as they appear in the buttons on the bottom of your screen, including:

- Music.
- Movies.
- TV Shows.
- Audiobooks.
- Top Charts.

- Genius (recommendations based on previous purchases).
- Purchased (access to everything you already own).

You can also search anything your heart desires in any of these categories by simply typing it into the search bar at the top, and allowing iTunes to do the work in finding it.

iBooks

When you want to download books from the iBooks catalog, you must download the iBooks app first. When you go, you can check books out of the library for free, for a discounted rate or you can buy them to keep.

Amazon

If you prefer Amazon as your point of purchase, you can logon to the website, or download the app through the App Store, and purchase music, movies, television shows, books and more. Each of these items will be available through iTunes, once they are downloaded. If you purchased music through Amazon.com, it will load into your iTunes automatically. So if you buy an episode of your favorite television show from Amazon.com, it will download to the "Videos" app, which is a default app that comes with the iPhone.

In order to read eBooks from Amazon, you will have to download the Kindle App for iPhone from the App Store. It is free, and will allow you to access all of your Amazon eBooks in one place going forward.

If you have an Amazon Prime membership, which included Amazon Instant Video access, you will simply need to download that app through the app store. It is free, and will give you access to all of your free Amazon movies and material as a result directly on your iPhone.

Downloading YouTube or Other Videos to iPhone

YouTube is a Google service, and Apple is not a huge fan of crossing their capabilities with competitors, so you will have to find a way around their rift in order to download videos. Tap on your App Store icon and search for video download software. One preferred option is "Video downloader - download manager." Once you have downloaded it (it is free), launch the app and search YouTube as you usually would. Once you get to the video you want do download, hit play and you will be prompted to download the video to your device. Download the video and play it back as many times as you would like later.

How to Back up iPhone Contents

In addition to your iCloud, there is way to clone the contents of your iPhone 6 to ensure that you never lose your music, videos, images, artwork, playlists, as well as song ratings and play counts, there is CopyTrans (http://www.copytrans.com).

CopyTrans provides a download for your complete iTunes library that allows you to organize it on your PC for awesome filing organization, and as a great backup – should anything ever happen to your iPhone. Keep in mind, there are never too many ways that you can back up your phone, so explore the available technologies with an open mind.

CopyTrans has more than just music cloning capabilities, these geniuses have developed downloads to backup, or clone, your other components as well, including:

- Apps.
- Contacts.
- Calendar.
- Notes.
- Photos.

Each download is available separately (or can be bundled into a group for savings!) and allows you access to each segment using your mouse and keyboard. This is important when you want to adjust your contacts by editing, grouping or adding information that your iPhone screen and keyboard simply will not allow with as much ease.

In addition, it is all of your prized iPhone possessions in one place: On your hard drive, which is the perfect alternate storage space. Also, the pricing is inexpensive, considering the pain, suffering and exhausted hours you would spend requesting everyone's contact information again and inputting the data manually. There is no disc, as the software downloads automatically to your computer so you can use it right away.

iPhone 6 Tips and Tricks

As you begin to play with your fun, new device, you will start figuring out shortcuts and options that you never knew existed. This is the nature of the beast with electronics. No matter how many manuals you read, or how many tutorials you sit through, practice really makes perfect with these devices, and sometimes trial and error does not hurt. For a few tried and true tips on operating your iPhone 6, check out the options below.

How to use Family Sharing

The Family Sharing feature allows up to six family members can share their purchases from the iTunes Store, iBooks Store, and App Store. This unique feature will allow your family to make purchases with the same credit card. Plus, you can approve kids' spending right from your device. In addition, the six family members can share photos, a family calendar, and more.

Close

Family Sharing

Family Sharing is the easy way to share what's important with members of your family.

Get Started

 Share purchased music, movies, books, and eligible apps.

 Share photos and videos in a family photo stream.

 Share your location with family members.

 Schedule events on the family calendar.

 Help family members find their missing devices

The family should designate one family organizer to set up this feature. When a family member makes a purchase, it is billed directly to the family organizer's account. To use this feature complete the following steps:

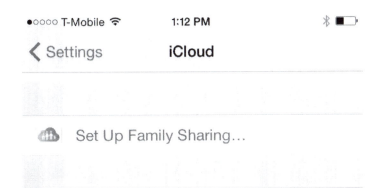

Set Up Family Sharing…

1. Go to Settings, tap iCloud, and tap Set Up Family Sharing.
2. Select which account will be the billing account.
3. Select payment method.
4. Choose to either "Share Your Location" or "Not Now" depending on if you want to share your location with your family.
5. Tap "Add Family Member" and enter the family member's name or email address. Repeat this step for all family members.

If you need to create an Apple ID for a child in your family, tap "Create an Apple ID for a Child" and enter the information required.

How to Connect with Other Apple Devices

When you use your phone at work, or with your small business, and need everyone to be on the same page, you can configure your iPhone to network with your other devices, including your iPad, MacBook or Mac Mini.

Network your mail, to do lists, calendars, contacts and reminders with just a few touches. You can also download apps like Dropbox that allow you to share information with other people in your company quickly and efficiently.

When you want to connect to your Mac computer, look for helpful apps like "Connect my Mac." However, you can also create a network between your devices, to access all of the items you need, even when you are on the go. To do so, create a network on your Mac, and share it with your iPhone 6.

How to Use Instant Hotspot

The great thing about your iPhone 6 is that you can use it as a hotspot to provide Wi-Fi access to your other iOS devices and Mac computers that are signed into the iCloud using the same Apple ID.

To use Instant Hotspot complete the following steps:

1. Go to Settings.
2. Tap Wi-Fi on your other iOS device, and simply choose your iPhone network under Personal Hotspots.
3. On your Mac or other iOS device, choose your iPhone network from your Wi-Fi settings.

Note: *This feature may not be available from all* carriers. Additional fees may apply.

How to use Personal Hotspot

Your iPhone 6 can also be used as a hotspot for your non-iOS devices. For instance, if you have a Kindle or other type of tablet, or if you have a pc laptop, you can still use your iPhone as a hotspot. This feature can be accessed as Wi-Fi, USB, or Bluetooth.

To activate Personal Hotspot complete the following:

1. Go to Settings

2. Tap Cellular, then tap Personal Hotspot (if it appears) and set up the service with your carrier.

After you turn on your Personal Hotspot, your other devices can connect to your iPhone 6 in the following ways:

Wi-Fi: On the device, choose your iPhone from the list of available Wi-Fi networks.

USB: Connect iPhone to your computer using the cable that came with it. In your computer's Network preferences, choose iPhone and configure the network settings.

Bluetooth: On iPhone, go to Settings > Bluetooth, then turn on Bluetooth. To pair and connect iPhone with your Bluetooth device, refer to the documentation that came with your device.

Note: *This feature may not be available from all carriers. Additional fees may apply.*

How to Extend Battery Life

There are a number of ways to extend your iPhone 6-battery life, each of which is a great idea – especially if you do not have a car charger!

Adjust Screen Brightness

Choose the lowest level that does not make you squint, but allows you to read the screen just fine. To do so, tap on the settings icon, and "brightness & wallpaper" on the left-hand toolbar, slide the brightness bar to your desired setting.

Keep Operating System Updated

Most iPhone 6 and 6 Plus devices are likely to have the iOS 8 software already installed. You should always check for an update from time to time, to make sure you are using the latest and greatest so your device runs optimally. You will be alerted when updates are ready, but it never hurts to check in case you breezed over the alert at some point.

Go to Settings > General > Software Update to manually check you have the latest version and if there is a new version you should be able to choose to have it installed to update your phone.

Additional Battery Life Tips

Turn off Line of Sight Motion: One quick tip to help with saving battery life is to turn off the background feature called "parallax." This feature is what creates the special background and foreground motion that changes with your line of sight perspective.

To shut if off, go to Settings > General > Accessibility > Reduce Motion and slide the button "On."

Turn off the Bluetooth: Settings > Bluetooth > Slide to Off

Turn off Notifications: Settings > Notifications > Turn Off All (or all you do not need to know about, like Scores, News, etc. when you really need to save battery power)

Use Wi-Fi, Not 4G: Settings > Wi-Fi > Slide to On > Enter Network Key

Set Your Sleep Timer: Settings > General > Auto-Lock > Select One Minute

How to Jailbreak/Unlock iPhone

When you "jailbreak" or unlock an iPhone, it is for the purpose of running software that is not approved by Apple. Depending on where you live, the legality of this process could be in question. However, there are a number of reasons why people do it.

The largest reason was to use ANY carrier, instead of only Verizon or AT&T. With an unlocked phone, individuals could take their service anywhere, with the luxury of still having an iPhone. Now that more carriers have signed on to the device's use, the next biggest reason became third party apps. This means the user is no longer tied to the apps that are available through the App Store, or that are not Apple approved.

The next most popular reason is to completely customize your iPhone. As Apple puts standards on the phone's settings, when it is unlocked users can apply designs, color schemes and user interfaces as they see fit. Of course, as with all devices that can be manipulated, individuals can also illegally download games, apps and information that usually requires payment for free, which is illegal no matter how you spin it.

How to Install Android on an iPhone

You can install Android on an iPhone, but it requires you to jailbreak your iPhone first, which completely voids the warranty – assuming you purchased one. Although the jailbreak process is relatively easy, it is to be practiced at your own peril. Although none of the common "jailbreak" methods have been shown to harm the phone and its overall functionality, you have to be responsible for any glitches that may occur along the way. In other words, it is not this guide's fault if you do so.

With that said, there are three popular jailbreak options for iPhone: PwnageTool, Redsn0w, and Blackra1n. Although there are a number of instructions listed online that will walk you through the process, but they all revolve around the following directions:

1. Once the iPhone is Jailbroken, run Bootlace in Cydia.
2. Install Bootlace by Tapping Manage > Sources > Edit and Enter: repo.neonkoala.co.uk.
3. Run Bootlace, and reboot the iPhone.
4. Return to Bootlace, and install OpeniBoot.
5. OpeniBoot, install iDroid.

This is going to take a while, so just sit the phone down and let it do its thing. Once it is complete, restart the phone one last time and the transformation should be complete!

It is worth considering, however, that if you purchased an iPhone 6, you probably did so because you are a fan of the technology. If that is the case, stick with the manufacturer settings and enjoy the phone as is. If you are an Android user who became vulnerable to the pressures of loyal Apple users, you may find it difficult to make the transition at first, but you will get there. If not, you can sell the iPhone 6 and buy a perfectly awesome Android of your choice with the money, or jailbreak the phone you have and just use the Apple shell, which sort of seems unfair to both technologies.

How to Set Reminders

You can make a to-do list of sorts using Reminders on iPhone 6. The great thing is, you can share these Reminders among all your Apple devices using the iCloud.

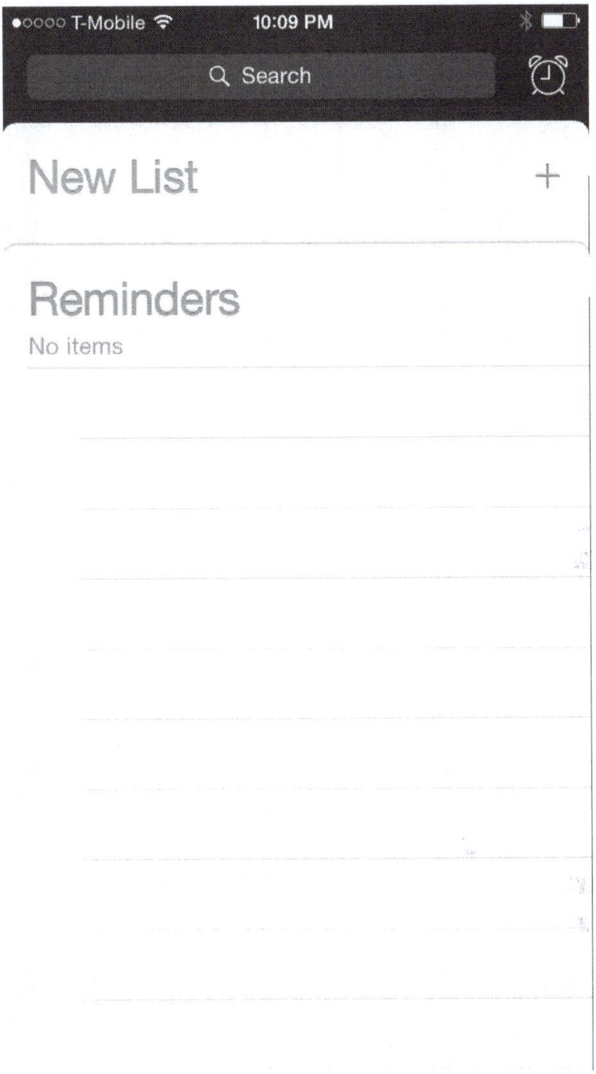

Q Search   ⏰

New List   +

Reminders

No items

One thing that makes Reminders better than a pen and paper list is that you can set them so that they are location triggered, which means your iPhone 6 will remind you of specific things you need to do when you are at a specific location. Amazing!

Create Reminders for Specific Days

Use the following steps to create a new reminder for a specific day:

1. Tap the Reminders icon on the Home Screen.
2. Tap the Add button.
3. A new reminder is started and the keyboard opens.
4. Type in the reminder.
5. Tap the Information icon to the right of the new reminder.
6. Tap the "Remind Me on a Day" switch.
7. Tap the "Date and Time" field.
8. Enter the date and time you want to be reminded.
9. If the reminder should repeat, tap "Repeat," and choose the amount of time before the reminder will reset.
10. Tap "Done" to save reminder.

Create new reminders for a certain place

Use the following steps to create a new reminder for when you arrive at or leave a certain location:

1. Tap the Reminders icon on the Home Screen.
2. Tap the Add button.
3. Type the reminder when the keyboard opens.
4. Tap the Information icon on the right of the entry.
5. Tap the "Remind Me at a Location" switch.
6. Tap the "Location" field and type the location's address.
7. Tap "Arrive" or "Leave."
8. Tap "Done" to save reminder.

How to Set Alarms

You may need an alarm to help wake you up in the morning, or perhaps you want to use an alarm for another reason. The good news is that it is easy to set an alarm on the iPhone 6.

To set an alarm, take the following steps:

1. Tap the Clock icon on the Home Screen.
2. Tap the "Alarm" tab at the bottom of the screen.
3. Tap the Add button at the top of the screen.
4. Set the alarm by scrolling through the numbers and AM and PM.
5. Tap "Save" to save alarm.

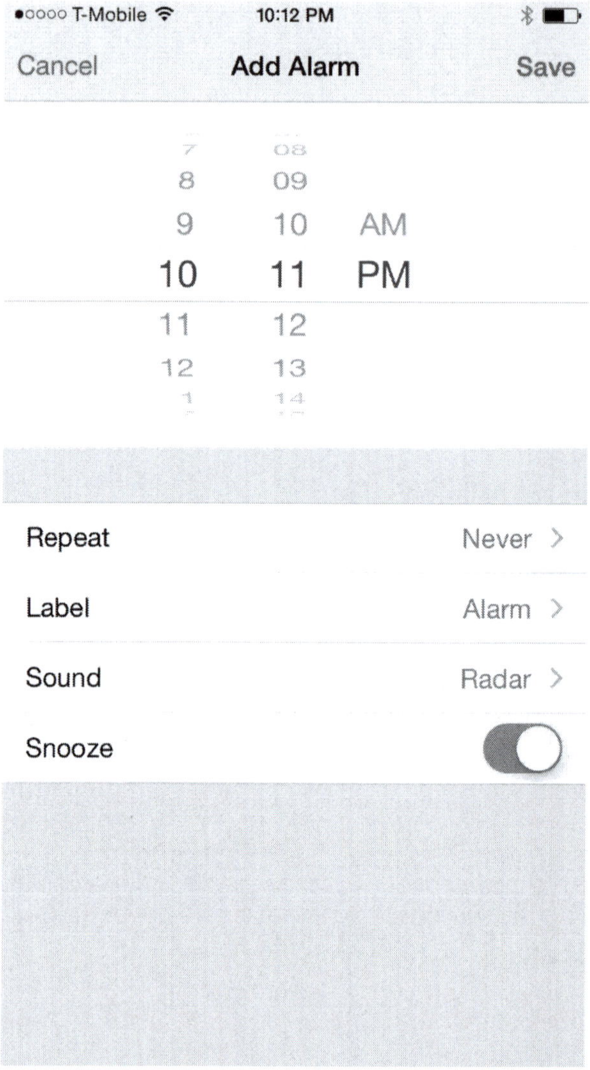

Cancel **Add Alarm** Save

7	08	
8	09	
9	10	AM
10	**11**	**PM**
11	12	
12	13	
1	14	

Repeat   Never ›

Label   Alarm ›

Sound   Radar ›

Snooze

If you want your alarm to repeat on multiple days or at the same time each week, do the following before tapping "Save:"

1. Tap "Repeat."
2. Tap the days you want the alarm to repeat, and a checkmark appears.

To customize your alarm sound and snooze do the following before tapping "Save:"

1. Tap "Sound." Choose the ringtone you want the alarm to sound.
2. Tap "Snooze" to turn it off or on depending on your preference.
3. Tap "Label" to name your alarm.

If you need an alarm at another time, simply repeat the steps listed above for the new alarm. When you have an alarm set, there will be a small clock showing across the top of your iPhone 6 near the battery.

How to Change Wallpaper

There's a great built-in setting that lets you easily change up the wallpaper on your iPhone 6 to your liking. With this feature, you can select from some of the pre-installed image themes already on the phone, or choose from those on your Photo Stream, Dropbox, Instagram and other locations.

To get to this feature, head to Settings > Brightness & Wallpaper. Once there, you can control the brightness of your iPhone, and also set Auto-Brightness to On or Off. You'll notice that your phone's lock screen and home screen are both on display here in smaller images. Tap on this and you'll move to the next screen that lets you choose from the pre-installed wallpaper options, or the images you have on your phone.

You'll be able to choose whether the wallpaper is used on your lock screen, home screen, or both and then preview the new wallpaper. On the preview screen you can also scale or move the wallpaper to your liking. Tap on "Set" if you want to use that as your current wallpaper and you'll see it on the home screen, lock screen or both!

How to use Airplane Mode

When you fly, you will need to put your iPhone 6 into Airplane Mode. The good thing is that while it turns off the phone's phone, e-mail, and Internet functions, you still have access to the phone's iPod functions.

✈	10:13 PM	▬▭

Settings

✈	Airplane Mode	⬤
📶	Wi-Fi	Off >
✳	Bluetooth	Off >
📡	Cellular	Airplane Mode >
📞	Carrier	T-Mobile >

📩	Notifications	>
🎛	Control Center	>
🌙	Do Not Disturb	>

⚙	General	>
AA	Display & Brightness	>
🌼	Wallpaper	>

To put your iPhone into Airplane Mode do the following:

1. Tap on "Settings."
2. Tap on "Airplane Mode."
3. Change the switch to "On."

To take your iPhone 6 out of Airplane Mode do the following:

1. Tap on "Settings."
2. Tap on "Airplane Mode."
3. Change the switch to "Off."

Once Airplane Mode is turned off, you will be able to make and receive calls and use the Internet.

How to use Weather for Different Cities

With your iPhone 6, you can track weather forecasts in cities around the world. Your iPhone 6 has a weather app that provides you with the current weather forecast for the city you are in as well as the cities of your choice. There is a six-day forecast for each city you choose.

To add a city to weather, do the following:

1. Tap the button in the bottom-right.
2. Tap the + button on the bottom of the list.
3. Type a city and state or zip code.
4. Tap Search.
5. Tap the name of the city in the search results.

Repeat these steps to add as many cities as you want.

To delete a city, do the following:

1. Tap the i button on the bottom right part of display.
2. Tap the red – button to the left of the city's name.
3. Tap the Delete button that appears to the right of the city's name.

When you are finished, tap "Done" on the top right of the display.

How to Get Time in Different Cities

The iPhone 6 will show you the current time in different cities around the world.

To check time in a certain city, do the following:

1. Tap on the World Clock–Time Zones icon from the Home screen.

2. New York and Berlin appear as the default "Favorites."
3. Add and delete your Favorite cities to your main World Clock screen.

How to Add a Third-Party Keyboard App

While the iPhone's on-screen keyboard is very good, some users may prefer to use a different keyboard, such as one that allows for "swiping" letters to form words. You can install a third-party keyboard app, with apps such as Swype, SwiftKey, Fleksy, Minuum and Adaptix.

The SwiftKey app is among one of the more highly recommended apps to use as it offers text predictability based on how you communicate. You can create an online SwiftKey Cloud account and it will learn how you write to help better predict what you will type. It also offers the "swipe" to type feature, where you slide your finger from letter to letter on the screen to form words.

Once you've installed a third-party keyboard (or a few), go to your Settings > General > Keyboards. Tap on "Keyboards" here to "Add New Keyboard." Select any of the new keyboard apps you want to add, and then make sure to "Allow Full Access."

Now, when you want to access any of your added keyboards, tap on the globe icon on your on-screen keyboard and you can cycle through your various keyboards to use.

How to Use the Emoji Keyboard feature

The Emoji keyboard is a feature many iPhone 6 users might not know about right away. This feature adds a new keyboard for sending messages that offers rows of "emoticons" – special faces, characters and symbols. It's important to note this will usually only display for other iPhone users or iPad users and not for many Android or other brands of smartphones.

To unlock Emoji keyboard, go to Settings > General > Keyboard > Keyboards > Add New Keyboard > Emoji.

To use this new keyboard, when you go to send text messages, you'll notice a small globe icon now, next to the microphone on your keyboard. Tap that globe icon once and you open the Emoji symbols that you can use with text messages. Tap the globe again to go back to the standard keyboard.

How to Activate Scientific Calculator

There's a basic calculator included when you tap on the "Calculator" app icon on your iPhone 6. Many users may not have realized that if you turn your iPhone sideways while the calculator is open, you'll get the calculator shown below, a full-fledged scientific calculator!

(	)	mc	m+	m-	mr	AC	+/−	%	÷
2nd	x^2	x^3	x^y	e^x	10^x	7	8	9	×
$\frac{1}{x}$	$\sqrt[2]{x}$	$\sqrt[3]{x}$	$\sqrt[y]{x}$	ln	$\log_{10}$	4	5	6	−
x!	sin	cos	tan	e	EE	1	2	3	+
Rad	sinh	cosh	tanh	π	Rand	0		.	=

How to Use Late Night Mode for Music

Late Night is a helpful feature to activate for listening to music when it's darker and quieter, usually towards nighttime, although it may have other times when it's helpful.

You can use a Late Night feature with the iPhone 6 built-in music app. Simply go to your Settings > Music > EQ > Late Night.

This will generally turn the bass down on your music and also reduce the tone somewhat for more appropriate nighttime listening.

How to Invert Colors on Apps (Night Mode)

There's also something like a Night Mode that is helpful for using apps. You can go to Settings > General > Accessibility > Invert Colors to On.

This particular setting will be particularly helpful for using a web browser at night, as it will invert colors (i.e. making white colors black, and black colors white). This feature may not look the best on all apps, so use it sparingly.

Bonus Tip: In the same area Settings > General > Accessibility you can also opt to switch on the Grayscale mode to make your iPhone have a black and white color scheme.

How to Use Do Not Disturb Mode

Enabling Do Not Disturb Mode can help to eliminate pesky sounds and alerts coming from your iPhone 6 when you don't want them to. You can even schedule the mode to occur during certain hours of the day of your choice, such as during the time you sleep, or another time of choice.

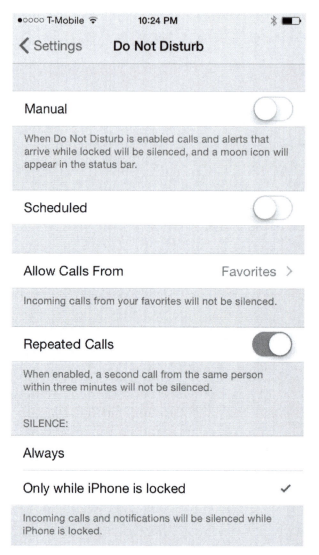

❮ Settings   Do Not Disturb

Manual

When Do Not Disturb is enabled calls and alerts that arrive while locked will be silenced, and a moon icon will appear in the status bar.

Scheduled

Allow Calls From   Favorites ❯

Incoming calls from your favorites will not be silenced.

Repeated Calls

When enabled, a second call from the same person within three minutes will not be silenced.

SILENCE:

Always

Only while iPhone is locked   ✓

Incoming calls and notifications will be silenced while iPhone is locked.

To activate Do Not Disturb Mode, go to Settings and move the Do Not Disturb button from OFF to ON.

To set up the Do Not Disturb mode timer, or choose which notifications, messages and calls you can receive or not, press on Settings > Notifications > Do Not Disturb.

How to Use the Find My iPhone

Let's face it, smartphones are nice and portable, but they can be easy to lose or misplace for some. With the Find My iPhone feature, you can give yourself a possible way to track down your phone, should this happen.

To activate Find My iPhone

1. Go to Settings.
2. Tap iCloud.
3. Tap Find My iPhone and slide toggle switch to On.

Activating this setting will let an iPhone owner track down their misplaced or lost phone using another iOS device. There are options to play a sound, enter "Lost Mode," which will allow the owner to set a passcode on the phone or send a message to the person who may have it. There's also the option to erase the missing device, should one want to prevent others from accessing what's on the iPhone. It's certainly one potential way to help locate a lost phone!

How to Monitor Data Usage

With various service and data plans offered based on which service provider your iPhone is with, you may want to keep an eye on how much data you have used on the iPhone. To do so, tap on the Settings button on your home screen. Next, tap on General, and then tap on Usage. Scroll down on that screen and tap "Cellular Usage."

You can now have a look at various cellular usage aspects on your iPhone 6 including "Call Time," and your "Cellular Network Data" as well as "Tether Data." If you want to track a new month or time period, you can also scroll down on that screen and tap on "Reset Statistics" which will set all stats to 0.

How to Take a Screenshot

When you want to take a screenshot of something that is currently in your view on the phone's screen, simple push the sleep/wake button at the top and the home button at the same time, and it will automatically save it to your camera roll.

How to Take Photos

Photo apps are a lot of fun, and can give you great options for filters, borders and backgrounds. You simply have to pick the one that is right for you by exploring the App Store's offerings.

Tap the App Store icon and search for "photo apps" to see the thousands of options available including Instagram, Pic Stitch, and Photo Editor – all of which are free to use.

Once they are downloaded to your phone, simply click on the icon you wish to use and take pictures just as you would without them. You will be using the same camera, just given more options to spruce up the image's extras.

How to Take a Photo While Shooting Video

The iPhone 6 offers options to take photo or video with the camera, but did you know you can take a photo as you're shooting a video?

To do so, while shooting a video on your iPhone 6, tap on the camera icon on the upper right hand of your screen. This will take a photo from the action while the video continues shooting. Photos will not be in full resolution, but will come out as 1920 x 1080.

If you want better quality photos, you can also make sure you are in photo only mode to take multiple photos as you are shooting video on the iPhone 6.

How to Bring Back Deleted Photos

Sometimes you might find you have accidentally deleted a photo you really love. If you have accidentally deleted a photo, it is possible to bring it back because there is a Recently Deleted album in the Photo app. This album stores deleted pictures for up to 30 days before they disappear for good. If it has been less than 30 days since you accidentally deleted a photo, you will be able to retrieve it from this Recently Deleted album.

How to Hide Photos

You can now hide a photo without having to delete it. To hide a photo, simply tap and hold a photo in the Photos app and an option to "hide" will show up. You will be able to remove the photo from Collections, Moments, and Years and keep it in the Hidden album. This is a great option to have if you have photos you would rather keep just for yourself.

How to use Timer for Selfies

Selfies are huge these days, and Apple has made it easier by adding a new timer to the Camera app. The timer allows you either 3 or 10 seconds to get the shot exactly how you want it. You can also prop up your phone before setting the timer, so you don't even need your hands to take a selfie.

To use this feature, open the Camera app to take a picture, select the clock icon, and set the timer. You can also use the timer with Burst Mode, which snaps a bunch at shots and you'll have plenty to pick from. Now you will always have the time to create the perfect selfie.

How to Control Camera, Music & Other Options on Locked iPhone

You can quickly access some of the features on your iPhone even without entering your pass code (if you've got one set).

On the lock screen, you will notice at the bottom of the screen there is a small camera icon in the lower right corner. Tap on this icon to quickly access your camera for taking a photo or recording video.

Additionally on the lock screen you can swipe up from the bottom center of the screen to bring up a further panel of options. Among these are Airplane Mode, Wi-Fi, Bluetooth, Do Not Disturb, the brightness control for your display, as well as a control for any music or video you are playing. Below that you'll find other features such as your phone's flashlight, timer, calculator and the camera. These are all available without having to enter a passcode or use your fingerprint scanner to access them.

How to Save a Web Image to Your iPhone

When you're browsing on the Internet, you can easily save photos from the web onto your iPhone 6. Simply press down on the image you want to save, and it will pop up with multiple options: Save Image, Open Image, and Open Image in New Tab.

If you press on Save Image, the photo should automatically be saved to your Camera Roll. To get to the photo go to your Home screen > Photos > Camera Roll.

How to Use Dropbox for File Transfers & Storage

The free Dropbox app is a great one to add to your iPhone 6, as it gives you an easy way to transfer files between devices. Not only that, but after you've signed up for the free Dropbox account, you can then log into your account on the internet anywhere you go and access your stored files.

Simply go to Dropbox.com to sign up for free for this service, and you'll get 2.5 GB of free storage space. Then, go to the iTunes App Store and search for "Dropbox." Install the free app on your iPhone 6. You'll also want to go and add the app to any other devices you might want to share files between (i.e. tablets, computers, or other smartphones in the home).

Dropbox even includes a setting, which will automatically upload your latest camera photos from the iPhone 6 to your Dropbox. This can be helpful for accessing photos easily that you took on your iPhone 6, but want to get on a computer, tablet or somewhere else online!

Create a Folder for Apps you Don't Use

A suggestion you may want to consider for freeing up app screen space is to create a folder of all the apps you don't use that can't be deleted. Whenever you press and hold down on an app icon on your screen, you'll see all apps start to wiggle. Apps with an "X" on them in a circle can be deleted, while those without the "X" can't. These include apps such as Nike + iPod, Calculator, Tips, Videos, Passbook, Weather, Notes and Health. You're stuck with these apps but can still free up space on your screen with this tip.

Make a folder to store these apps in. Press and hold down on any app icon on your screen that you want to store away. Next, drag it on top of another app that you don't want on your screen, but can't delete. Release your finger and a folder will be created. You'll be able to name it something like "Unused Apps." Continue to drag and drop other apps you don't use (and can't delete) to this folder, and you'll free up some extra app screen space. When you're finished, simply press the Home button on your iPhone and all the apps will stop wiggling.

How to Make Secure Credit Card Purchases

A cool new feature that comes with iOS 8 is a credit card scanner for making secure purchases. When you're making a purchase on the Safari browser that uses a credit card, you now have an option to "Scan Credit Card," instead of entering your card information. The feature will appear above your on-screen keyboard on any site where you're required to enter your card info. Once you select the option you'll simply hold your card up in front of your iPhone's camera lens for it to securely capture your card info.

How to Mute Text Message Conversations

Tap on any person's text message on your iPhone messages area and you'll see "Details" on the upper right hand corner of the message thread. Tap on "Details" and it brings up additional options including "Do Not Disturb." By sliding the button on, you can mute this specific conversation so it won't send you alerts or notifications. This can be helpful if you want to continue receiving the messages minus the alerts.

How to Delete or Share Text Messages

Once you're in a text message thread on your phone, you can press down on an individual message in the conversation to bring up "Copy | More..." Tap on the "More" option and you'll now see radio buttons next to each of the text messages. With this, you can tap on any radio button next to a message and delete that individual part of the conversation. Additionally, you can use this to share any specific message with another contact.

How to Create a Medical ID for Emergencies

With iOS 8 comes a new Health app with plenty of cool features that will be helpful for tracking health, fitness and nutrition goals. A special feature that this app includes is the ability to create a Medical ID. This is helpful medical information that someone could access from your phone in case of an emergency. The info would be accessed from the emergency dialer area of your iPhone so that it wouldn't be necessary to unlock your phone.

To create the Medical ID, go to the Health app and then tap on the "* Medical ID" option on the lower panel of the screen. You'll be able to enter your birthday, name, any medical conditions, medical notes, allergies or reactions, as well as an emergency contact, blood type and organ donor. Once you're finished entering information, tap on the "Done" option at the top of your screen to save it. Now you'll have this valuable info stored on your phone and accessible in case of any emergencies.

25 Free Apps to Improve Your iPhone 6

There are literally hundreds of thousands of apps available on the iTunes App Store, however not all of them are going to be helpful. Here's twenty unique, interesting and helpful apps that you may want to head over to get at the iTunes App Store. The best part about these apps is they are all free versions (with some offering paid versions as well!)

1Password – This app is a great way to store all of your passwords using one convenient and secure app. You simply set a Master Password that can be used to access your vault of passwords. In addition, the app has integration with the fingerprint sensor. It's important to note that the app is free, but to unlock more additional features you'll need to spend 99 cents.

Adobe Photoshop Express – This is a free photo editor app from one of the top names in the business. Use it to make light but effective photo edits on your iPhone or other mobile device.

Calorie Counter & Diet Tracker by MyFitnessPal – This is a great free app from MyFitnessPal to help you with your diet and calorie burning tracking needs. Simply sign up for free at their website and add the app to your iPhone, then use it to track your daily intake based on recommended foods, or search the database and add in the foods you've consumed to track calories. Select the various activities you've performed during the day to get a measure of how your calorie-burning goals are progressing.

Camera Awesome – Take your photos to the next level with special filters and features.

Chrome – An alternate browser to Safari, Chrome is made by Google. If you're a Chrome Internet user on your laptop or other device(s), you can sync bookmarks, passwords and other data to your iPhone.

Converter Plus – A great app for making conversions whether it's the American dollar to another form of currency, metric to imperial measurements, loan interest figures and more.

Dropbox – This app gives you 2.5GB of free storage. Use it to share files between your devices and online for easier access. There are also paid versions for more storage space.

EasilyDo – This is a productivity app at its best. EasilyDo can scan your various online services such as Facebook, emails, calendar and more. It will then look for tasks it might help you get accomplished, including wishing someone a happy birthday on Facebook or adding a new contact to your emails.

Evernote – A productivity app that is useful for organizing projects by individuals and business owners. Set up notebooks, add notes by text or voice, along with photos, web clippings, links and other essential details, and then sync with your other devices.

ESPN Score Center – Sports fans will love this free app as it lets you stay updated on the latest scores. Add in your favorite sports or teams and get the latest scores, results, news and other pertinent info delivered to your iPhone 6.

Flipboard – Make your own customized news magazine with the stories and content you want including Twitter, Instagram, Facebook and Google + posts along with the latest from around the web.

Google Maps – The iPhone 6 includes Apple Maps, but some prefer Google Maps. This free app will give you voice-guided-turn-by-turn directions, as well as directions to use any public transportation methods as well. In addition, you can see ratings and reviews of any restaurants or other venues you might be heading towards.

Google Translate – Going on a trip to a foreign country but don't quite speak the language? Google Translate is a free app that can help you with that issue, as well as help those who are looking to learn a language. Install the free app on your phone and be prepared for those instances you might get lost in translation.

Instagram – A popular social media app that allows you to share photos with cool effects.

Opak – This on-board photo app for your iPhone gives you access to "Instagram-style" filters as well as stickers and other add-ons to make your photos that much more interesting, fun and appealing.

Open Table – A great app that allows you to use Siri to make quick reservations at a wide number of eateries and restaurants.

Kayak – A helpful free travel app from the Kayak travel search and booking company. Use this app to find the best deal on airfare and book it with the app, making your travel plans that much easier to make!

Kindle – If you love to read, the Kindle app from Amazon is the way to go. This free app lets you read items you've purchased through Amazon such as eBooks, magazines and newspapers.

MyScript Stack – Handwriting Keyboard – While the on-board keyboard is great and so are third-party keyboards like SwiftKey, this app can be helpful in that it allows you to "type" by writing with your finger. This one may not be for everyone as it takes some getting used to, but can be fun to test out on your device!

Pinterest – This is a popular and free social media app that acts as a digital scrapbook. Pin the products, items and things of interest you find online to your boards and check out other people's latest finds.

Spotify – A free radio on the go app. With this app you can search for any artists, songs, albums or genres of music you like and create your own personalized playlists on the go. Playlists will sync with Spotify if you have it set up on your other devices as well.

TripIt – An extremely helpful travel app for those on the go who want one place to organize their itinerary. With TripIt you can put all your travel confirmations including hotel, airfare, car rental, restaurants and more to create one simple itinerary on your iPhone. You can also share the trip plans on your Facebook and sync it with your calendar for more accessibility.

UP by Jawbone – While the latest iOS update from Apple included a Health app, there are many great apps out there which integrate nicely with it. One of the best free ones is UP by Jawbone. With this app you can track your daily activity automatically to stay on top of your fitness and health goals. In addition, the app lets you manually enter sleep or diet data to further stay on top of things.

Vine – a video app brought by the makers of Twitter. With this app you can create six-second looping videos that have sound or no sound. The intuitive design allows you to put multiple parts of video footage you took into one looping video. Then, share it on your Vince, Facebook or other available platform.

Wunderlist – This is a great app for anyone looking to stay more organized, productive and on top of those daily, weekly or monthly tasks. Wunderlist helps you organize lists that are important to keep track of including shopping lists, "to do" tasks and much more. The latest free version even includes integration for Dropbox files.

10 Great Free Games for iPhone 6

The iPhone 6 isn't just about staying productive, making notes, sending emails, making calls or other day-to-day tasks, but it also allows you to fully unleash the impressive mobile, handheld gaming capabilities. Here are ten free games you might want to install on your iPhone 6 for downtime when you want to have some fun! (Keep in mind some of these selections also offer "in-app" purchases.)

Candy Crush Saga – One of the most popular games on Facebook as well as in the mobile community, in this game you match up various same-colored candies to remove them from the board, scoring points and clearing levels. There are various challenges, boosters and other fun items along the way as you move along the board.

Clash of Clans – An epic combat strategy game, Clash of Clans lets you build up a village, train troops and then battle with other clans competing online. There's a great social aspect to this game, along with in-app purchases, upgrade levels, and a campaign to fight against and defeat the Goblin King. Amazing graphics and entertainment value in a free game download!

Draw Something – This is a social game much like Pictionary yet for online players. Player one receives three different words and chooses one to draw. The player beings to draw it on the screen and has no time limit. Player two will log in and see the image as player one is drawing it, and then takes guesses based on the number of letters in the word.

Dumb Ways to Die – A free game that has caught on in terms of popularity. In this one you are performing various difficult tasks with a time limit. They might include "swat the wasps" or "slide the fork out of the toaster." Of course, as you move along, the tasks get harder, and you just might run out of lives if you can't avoid one of those dumb ways to die!

Fast And Furious: The Game – There have been six movies and counting and now a free game to go with them. In this racing game you can also play in "Heist" mode to escape the police after a bank robbery, customize your rides, and even join crews with friends to enter tournaments. It's free, but there are all sorts of in-app purchases to add to your gameplay too.

Flow Free – A highly addictive free puzzle game. In this one you are solving the levels by trying to connect same-colored dots by tracing your finger around a grid. The object is to fill in every square on the grid with a line, while not crossing lines. It's one you have to see and experience, but once you do you'll love it.

Fruit Ninja Free – A free version of the hit game Fruit Ninja. With this one you're using your finger to swipe across the screen and slash fruit as it shows up, as quick as possible!

Layton Brothers: Mystery Room – If you enjoy shows like CSI or Law & Order, this may be the game for you. Put on your private investigator cap because in this free game you're given a set of clues to find out what happened in a case before someone gets prosecuted. Made from the creators of the Professor Layton series for Nintendo game consoles.

NY Times Crosswords – This one's free, but you can also choose to purchase a $2.99 per month subscription for it. The New York Times is considered the king of crossword puzzles, and this free app gives you a limited selection of puzzles to solve, helping to stretch your brain by learning new words and facts along the way.

Temple Run 2 – Temple Run was a wildly popular game that involved an adventurous character dashing through the jungle. Along the way, you've got to avoid obstacles by turning corners, jumping and swinging. This free sequel to the original has become just as popular and will be yet another addictive game on your iPhone!

Troubleshooting iPhone 6

Occasionally, you will experience some issues with your iPhone 6. The following sections have some troubleshooting solutions for your iPhone 6.

Rebooting your iPhone 6 (Frozen, Crashed or Locked-up Phone)

There will be times when you simply just cannot maneuver through your phone as you would like. Just like any computer – which this small device actually is, thanks to its powerful operating system – it may frustrate you to pieces at times. This includes a frozen app, a blank screen or even the inability to maneuver between your onscreen app's selections. When this happens, rebooting the device may help.

This rebooting process is just a forced shutdown that should not affect the integrity of your operating capabilities, apps, music or contacts in any way. You are not going to damage your iPhone by rebooting it. It is simply the equivalent of holding in the power button on your PC until it restarts itself, which everyone has done at least one thousand times in their lives. If you never have to use this function, good for you! If you do have to, you will be glad you know how.

- Hold the sleep/wake button at the top right of the device and the Home button at the bottom center of the face at the same time.
- Hold both until the screen goes black – or up to ten seconds. **NOTE:** *If the red slider appears that says, "Slide to power off" ignore it and keep holding the buttons simultaneously.*
- When the silver Apple logo appears, let go of the buttons.
- Rebooting is complete. You should be able to use your iPhone with no trouble!

Re-setting your iPhone 6

If you are having tremendous trouble with the operating system, or are planning on selling the iPhone or transferring its ownership to another, you are going to want to restore the device to its manufacturer settings. This will ensure that all of your personal information, music, contacts and apps are deleted for good, so you can rest easy at night knowing that the person who bought it on eBay is not enjoying their daily latte using your virtual wallet.

Keep in mind that if you are doing this to fix a problem, and have already followed every other troubleshooting possibility that you will need to back the phone up first so you can restore your contacts, and everything else once reset it for good.

When you are ready to reset the device, follow these easy instructions:

- Launch "Settings" and tap on "General."
- Scroll to the bottom of General and tap on "Reset."
- Tap on "Erase All Content and Settings."
- Enter your password, if necessary and confirm the reset by tapping on "Erase iPhone."

This process will restore the phone to its factory settings, so you are now free to get rid of it or to reload your data accordingly. It will also take a while to accomplish, rendering the phone useless while it proceeds with the complete reset. Once it has finished, the iPhone will reboot and revert to the familiar factory setting screens that you completed when you first received the phone. At this juncture you have three options:

- Leave it as is for the next person (buyer or gift recipient) to set up on their own.
- Reprogram the device as a brand new setup for yourself.

- Restore the device from your backup files.

iPhone 6 Accessories to Consider

Your enjoyment of the iPhone 6 or 6 Plus should not be limited to just what's in the box. There are plenty of additional accessories on the market for the mobile phones, with more being developed or unveiled as you read this. You can maximize the value of your item with these upgrades for storage, protection or entertainment.

Wireless Bluetooth Speaker

When you carry everything on your new iPhone 6, including your music, you are going to want to access that entertainment at all times. As the world begins to lean towards wireless technology, so do entertainment features, including how you listen to music.

Bluetooth, or wireless, speakers allow you to access them remotely, so you can play music as a party or at home, without plugging your iPhone directly into the device. This is a great feature when you do not want to leave your phone behind in a strange place, but still want to access your music library.

Bose, Beats by Dr. Dre, and even the lesser-known Sono brand, all have wireless speakers that you allow you to access them with your iPhone 6. Simply turn the speaker on, tap your setting icon, tap Bluetooth (make sure it is on) and it will locate the wireless device. Tap on the option, which is usually the name of the device, to connect. This will take no time at all.

Chromecast Dongle

The Chromecast Dongle is a new piece of technology introduced by Google in 2013. It's as small as a USB jump drive and plugs into the HDMI input on your hi-definition TV. Once it's plugged in and set up you can stream content from your iPhone to the TV. As of this publication, it works with Netflix, YouTube and Google Play content, although more apps are being added for various mobile devices.

Lightning to 30-pin adapter

With this cable, owners of old iPhone accessories can still use them! The cable will support USB audio, analog audio out and syncing. However there is no support for iPod out, a feature that works with certain car stereos to control the phone.

Lightning Dock

Consider the lightning dock, an aluminum dock that lets you connect your iPhone whether it's got a case on it or not. This accessory allows your iPhone 6 to stand upright on display and will charge or sync your phone with your lightning cable and a computer. The dock is said to also work with other devices including iPods and iPads so it could make a great investment if you have multiple Apple devices to dock, charge and sync. The Lightning Dock for previous iPhone models sold for $40 or less. Make sure that the model you purchase will support the dimensions of the iPhone 6.

iPhone Car Mount

If you're on the go quite a bit and don't want to fumble around with your iPhone, then a car mount is the perfect solution. One recommended mount to take a look at is by iOttie, which provided options for the iPhone 5 models. This is a well-designed car mount that will lock your iPhone into place and keep it mounted on the dash for easy access during travel. Simply use the push of a finger to unlock the phone and pop it into your hand. Includes 360-degree rotation to display your phone in portrait or landscape mode on the dash. As with other accessories, make sure to choose an item that will work with the dimensions of the iPhone 6 or 6 Plus.

Stylus Pen

A lot of tablet and mobile phone owners use stylus pens as a better way to navigate, tap and swipe on their screen without smudging up the display. A good stylus pen such as the Kuel H14 series can help keep your display visible and make overall touch on your screen smoother and smudge-free!

High Capacity Power Bank

A power bank will give you extended battery life for your iPhone when you're on the road, whether it be a long trip by car, plane, train or a cruise on the seas. One good power bank is the series by Anker Astro, as these devices include two power inputs on the front of the display, making it ideal to charge an iPhone and another device at the same time. Use it on that next long road trip to keep your phone alive and extend the time you can talk or use the device. These generally sell for under $50 depending on the capacity you buy, but keep in mind they'll weigh a bit and add to your accessories for packing.

Another option is the MoJo Hi5 Powerbank which costs about $80 or less and attaches around your phone. The case includes smart features to make sure you don't overcharge the phone or short circuit it while charging. This could be a more portable and convenient solution but may not provide quite the extra power as a bigger power bank.

Wi-Fi Hard Drive

If you need more space than your iPhone 6 or iCloud offer, or simply want to keep all of your content in one place where it can be accessed by your other devices, a Wi-Fi hard drive is a great idea.

This will also let you carry along a library of movies if you are traveling with children or let you store pictures as you take them with your device, for a wedding, reunion or work related event. Look for the trusted names in Seagate and Kingston or Patriot and enjoy up to 64 Gigabytes of extra space on a Wi-Fi network with no problem.

In-Car Bluetooth FM Transmitter

An in-car Bluetooth FM transmitter like the GOgroove FlexSMART X2 would be a great addition to your iPhone 6. This accessory allows for hands free calling and music control while driving. This device can also charge your iPhone 6 while you are streaming music to it.

When you receive or make a call, the music fades out, and its enhanced voice detection provides for superior call quality. Plus, you have the ease of making hands free calls while you keep your hands on the wheel.

Upgraded Headphones

There are so many great headphones available, and it can be hard to choose. An excellent choice in headphones to go with the iPhone 6 is the iLuv ReF Headphones. These headphones have canvas covering the ears, which is somewhat unusual and provides a unique style flavor in comparison to other over ear headphones.

In addition, they provide excellent sound quality, and the collapse, so that you can easily fold the up and take them with you wherever you go. Plus, there is a hands free mic, so that you will not miss a call while enjoying music or listening to an audiobook on your iPhone 6.

If these don't suit you, there are plenty of other great headphones like Beats Solo HD On-Ear Headphone or Monster NCredible NTune.

Sports Armband

If you run or do other exercise while listening to your music on your iPhone 6, you know it can be hard to carry around. A good quality, waterproof sports armband is a must for people who exercise often with their phones.

A great option is the Belkin ProFit Armband. This band adjusts once, and then you can easily strap it on and off without readjusting each time. It is also waterproof, and comes in a variety of colors.

Apple Watch

In early 2014, Apple will release their very first smartwatch called the Apple Watch. This smart device will be able to integrate with your iPhone in all sorts of amazing ways. The watch will bring notifications of new mail, messages and calls, as well as important updates from mobile apps. In addition, it's expected to offer features that work with the new health apps, as it offers a custom heart rate sensor and many more helpful features. Even more, the special "Taptic Engine" that this smartwatch uses will give different sensations on your wrist coupled with subtle sounds for different notifications, alerts or features. See more info at Apple.com for this exciting new accessory.

Where to get More Help with iPhone

If Apple has made one thing clear, it is that they like to be the troubleshooting resource for all of their products. Because of this shared mentality within the technological giant, the Apple website is a great source of information for product Q&A, troubleshooting guides and support.

Within their website they have online Apple Forums and Online Communities directly on their website, where you can submit a question and receive answers from Apple device users around the world. This is a helpful tool because there are a lot of different experiences that can be shared in one place, and the results are explained by a real person, who is not necessarily speaking tech talk and confusing you even further.

If the online communities with Apple.com are not providing you with the answers you need, simply locate the "Support" tab at the top of their website, choose the iPhone product when prompted, and search through any of the following resources for help:

- Videos.
- Manuals.
- Tech Specs.
- Online Support.
- Telephone Support.

When you need help now, and respond better to in the flesh support, you can visit a genius bar at your local Apple store. Genius bars are in-store help from real employees who know Apple products inside and out. They can offer advice, quick fixes and even suggestions for fixing your phone whether it is under warranty or not.

To set up an appointment online, logon to their concierge service at the following URL: http://concierge.apple.com/reservation/us/en/techsupport/

Select your state, and the store you would like to make the appointment in from the drop down menus. Next, pick your device – in this case the iPhone. It will ask you to reboot your iPhone, to see if that resolves your issue. It will also give you the opportunity to be re-routed to their online resources, or you can simply click "Continue with reservation" to get to the timeslots available.

It is HIGHLY recommended that you make an appointment at the Genius Bar, and do not simply show up at the Apple Store instead. Everyone (well, most) respects the system, and the Genius's time and knows that plowing through the line with your device will get you summarily rejected by the Genius you approach, and the scores of individuals politely (well, most) awaiting their turn. If you cannot make an appointment online, call the store nearest you to set one up over the phone. You will be glad you did.

Outside of the actual Apple sanctioned resources, there are thousands more online that can help you get through troubleshooting techniques that Apple may or may not approve of. Some of them are the same as what the folks at Apple will tell you, but maybe delivered in an easier to understand explanation.

Use your favorite search engine and ask the iPhone related question that is plaguing you. You will receive answers from everywhere, including Apple, about.com, Amazon reviews, forums, online communities, and iPhone retailers who are simply looking to sell you accessories.

The good news is, no matter what problem you are having, there is probably another Apple user who has experienced it too, and shared it with the world. This is what technology brings to the table: A worldwide forum of Apple users who can solve each other's problems online!

In all seriousness, look for help from a qualified and friendly Apple employee first. They are really great at walking you through steps over the phone or in an online chat. If all else fails, lean on the Mac world to give you a few pointers here and there. You never know, you may just learn something new about your iPhone that you would not have known otherwise, and that is what it is all about!

Conclusion

The iPhone 6 and 6 Plus are currently among the top choices on the smartphone market by many consumers. Apple tends to offer plenty of great specs and features on their devices, which when understood, can really aid in comfort, day-to-day productivity, time management and important situations. Phones have come a long way since the landline telephones, and now with a device that has all these capabilities, your life can be easier if you learn how to properly use your phone and take advantage of its capabilities.

Expect Apple to release important iOS updates for the iPhone 6 in the future. While iOS 8 was released in 2014, more updates will arrive to provide fixes for bugs. In addition, each new update offers more great features and tools to really take the potential of your iPhone to the next level!

More Books from Tech Media Source

Apple TV User's Guide: Streaming Media Manual with Tips & Tricks

iPad Mini User's Guide: Simple Tips and Tricks to Unleash the Power of your Tablet!

iPhone 5 (5C & 5S) User's Manual: Tips and Tricks to Unleash the Power of Your Smartphone!

Kindle Fire HDX & HD User's Guide Book: Unleash the Power of Your Tablet!

Kindle Paperwhite User's Manual: Guide to Enjoying your E-reader!

How to Get Rid of Cable TV & Save Money: Watch Digital TV & Live Stream Online Media

Chromebook User Manual: Guide for Chrome OS Apps, Tips & Tricks!

Chromecast Dongle User Manual: Guide to Stream to Your TV (w/Extra Tips & Tricks!)

Google Nexus 7 User's Manual: Tablet Guide Book with Tips & Tricks!

Samsung Galaxy S5 User Manual: Tips & Tricks Guide for Your Phone!

Samsung Galaxy Tab 4 User Manual: Tips & Tricks Guide for Your Tablet!

Amazon Fire TV User Manual: Guide to Unleash Your Streaming Media Device

Roku User Manual Guide: Private Channels List, Tips & Tricks

Printed in Great Britain
by Amazon.co.uk, Ltd.,
Marston Gate.